ALSO BY MAVIS AMUNDSON

Sturdy Folk
The Lady of the Lake

THE GREAT FORKS FIRE

By

Mavis Amundson

Western Gull Publishing, a division of
the *Peninsula Daily News*

Copyright 2003 by Mavis Amundson
All rights reserved, including the right of
reproduction in whole or in any part in any form.

Western Gull Publishing
A division of the
Peninsula Daily News
305 W. First St.
Port Angeles, WA 98362

ISBN 0-9610910-7-X

Olympic Printers, Inc., Port Angeles, Seattle, WA

*For my father,
Ben Dodge,
who drove Caterpillar tractors
with the best of them, and
for George.*

Contents

INTRODUCTION

CHAPTER 1 *"It's a Fire"* 1

CHAPTER 2 *A Dry Summer* 5

CHAPTER 3 *Frontier Spirit* 16

CHAPTER 4 *The Fire* 21

CHAPTER 5 *Forks Fights Back* 28

CHAPTER 6 *Epilogue* 37

ACKNOWLEDGMENTS 45

NOTES 48

SOURCES 53

INDEX 55

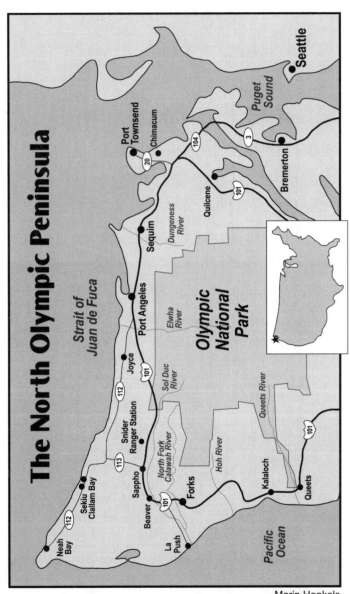

Introduction

On a fall day in the early 1950s, a fast-moving forest fire roared across the North Olympic Peninsula in Washington State and headed toward the timber town of Forks, creating a scene of almost terrifying proportions.

The blaze raced about 18 miles in less than a day, blackening a swath of forest up to five miles wide, some reports say. Flames leaped from treetop to treetop. Strong winds flung glowing embers a half mile ahead of the main blaze.

Smoke billowed thousands of feet into the sky and obscured the sun, creating an eerie twilight. Burnt needles and small branches, carried aloft by powerful winds, pelted rooftops and yards in Forks.

As the fire bore down on the timber town, authorities ordered Forks evacuated. But many people stayed behind, determined to save their town.

The fire made everyday heroes out of men and women who grabbed hoses and watered down houses and businesses. Others risked their lives on fire lines. Some made sandwiches and coffee for the people on the front lines. Many men and women scrambled to save their loved ones and their most precious belongings.

Loggers, sawmill workers, firefighters, shopkeepers, retirees and others helped save their town from the advancing fire. With little more than bulldozers, shovels, garden hoses, some outdated firefighting equipment and a pitiful water supply, these heroic townspeople helped hold back the forces of nature.

Their effort was extraordinary, but not unexpected. Like their pioneer forebears, who carved out a living in this beautiful but inhospitable wilderness, the people of Forks were imbued with the frontier spirit of toughness and self-reliance. As the fire encroached upon their town, they fought back. They didn't give up. They had families and homes to protect.

They took care of their own.

CHAPTER 1

"It's a Fire"

Ella Brager was up early. A traditional homemaker of the 1950s, Ella liked to fix breakfast for her husband, Clarence. He was a logger, and already he was up and outside getting his rigs ready for a day in the woods. The young couple's two daughters, 11-year-old Suzanne and nine-year-old Nancy, were asleep in their beds.

It was September 20, 1951, a typical September morning in the Brager household, or so it seemed.

But Ella was uneasy. Something was wrong. The ringing of the telephone had awakened her sometime after 3 a.m. The phone call was for Clarence's brother, Lawrence, who shared the Bragers' two-party telephone line. "That's odd," thought Ella. Who would be calling her brother-in-law in the middle of the night?

The Bragers lived in a modest wood-frame house in Forks, a timber town of about 1,100 in the northwest corner of Washington State. The town's livelihood depended on the hemlock, spruce, cedar and Douglas fir that grew

1

in abundance on the slopes and valleys of the Olympic Mountains.

Clarence and his brother ran Brager Brothers, a logging outfit that operated in the Olympic forests. Their crews went out each morning, logged the big trees in the woods, and hauled the logs to nearby landings. Ella worked as the company bookkeeper and frequently ran a cookhouse to feed the loggers who often bunked on the couple's two-and-a-half-acre spread.

"Look out here," Clarence said to Ella as he came back into the house. The news from his brother, Lawrence, was not good. Clarence gestured to the east, toward the Calawah River Valley, a heavily forested valley in the Olympic Mountains.

"It's a fire," Clarence said.

Ella took a few steps from the kitchen to the back porch. From there she looked out at the valley. It was before dawn, yet something resembling a sunrise appeared on the eastern horizon. The sky was aglow, a vivid red.

Clarence was right. It was a fire.

Ella, a brave logger's wife, was fearful for her husband. She knew what Clarence had to do. The company kept its logging equipment at a worksite in the woods; logs were stacked to be hauled away to markets. Ella knew that her husband had to go into the forest, perhaps in the path of the fire, to save the company's logging equipment. It was risky, she knew that, but she also knew it was necessary.

Clarence and Ella Brager and their daughters, Nancy and Suzanne.

Fire danger was part of their lives. A decade before, a fire in the woods had come perilously close to the Bragers' first home 10 miles south of Forks. Sparks from the fire landed on a tree in the Bragers' backyard and set the tree on fire. Ella grabbed up her baby daughter and fled to safety. But the fire was put out quickly before it reached the Brager home.

Today was different. Already, a strong east wind was driving the fire at a fierce rate of speed, with a wind so strong that it carried burnt fir needles to Ella's doorstep. She felt the burnt needles raining down, and she watched her dog brush the needles with his tail – remnants of the blaze miles away.

By mid-morning Ella scrambled to save her family and their belongings. "I don't think anyone panicked," she recalled. "We just knew we'd better get going." She asked herself, "What should we take? What is most important?" Neighbors congregated in the street asking the same questions.

Clarence returned home from the woods safely by late morning. He came into the house where Ella was packing. "You'd better get out of here," he said.

The fast-moving blaze was threatening Forks.

CHAPTER 2

A Dry Summer

The summer of 1951 was unusually dry on the North Olympic Peninsula, where the Calawah River flows. The region is bordered by Hood Canal, the Strait of Juan de Fuca and the Pacific Ocean. The heart of the Peninsula is Mount Olympus and the Olympic range, a rugged landscape that is the origin for several major rivers, including the Hoh, the Sol Duc and the Bogachiel, whose tributaries include the Calawah.

The Calawah River originates at the confluence of two small streams in the Olympic foothills. The river then flows west and south, picking up tributaries along the way, through a forested valley that opens up at the Forks Prairie. There, settlers founded the town of Forks on an expanse of relatively flat land sandwiched between the south bank of the Calawah and the western foothills of the Olympics.

The region is famous for its heavy rainfall, but the town and the Calawah River Valley were parched in 1951. U.S. forestry officials at the Soleduck Ranger District, which managed the area's federal forests, noted that the spring and summer of that year were among the driest on record. It was so dry in 1951 that officials closed the Sol Doc River area between July 2 and September 15 because of extra fire danger.

Similar forest conditions were reported throughout the Northwest. The region's timberlands were tinderboxes. Fire danger was high.

Dozens of forest fires broke out that fall in Washington, Oregon, British Columbia and Northern California. One blaze swept over 2,000 acres of mostly forestland in British Columbia and destroyed logging equipment. In Oregon, a flash fire covered 600 acres and threatened a logging camp. Fires blazed out of control in Northern California.

Fires took a devastating and sometimes tragic toll. Just two years before, 13 smoke jumpers were killed in an out-of-control fire in the Mann Gulch wilderness in Western Montana. Many people still remembered the Tillamook Burn in Oregon, whose main blaze in 1933 raged over a quarter-million acres.

On the Olympic Peninsula, loggers and timber companies took fire precautions but kept going to work in the woods. Among the landowners harvesting trees in the area were Merrill & Ring, Rayonier Inc., Fibreboard Products Inc., and the U.S. Forest Service. In addition, independent loggers, called "gyppos," like the Brager Brothers, contracted with companies like Rayonier and sent their own crews out into the woods.

Still, spot fires bedeviled woods workers through that dry September. The Calawah River Valley, surrounded by heavily forested hills and steep terrain, was especially vulnerable to fire. Recent logging activity left parts of the valley covered with dry, flammable wood debris.

A particularly stubborn fire broke out in early August along the Port Angeles and Western Railroad right-of-way near Camp Creek, not far from Lake Crescent and about 18 miles east of Forks. The fire was only one in a series of problems besetting the troubled railroad.

The railroad had lost much of the luster of its early years, when, dubbed the Spruce Division Railroad, its mission was to haul Sitka spruce logs from Olympic forests for use in airplane production during World War I. But the war ended before the spruce railway project got off the ground, and afterwards the railroad changed hands and eventually fell on hard times.

The Great Forks Fire

Financial difficulties forced the railroad to cut back on operations, including making scarce use of "speeder cars," which ordinarily followed trains to detect fires. Instead, the railroad patrolled for fires by routinely retracing its runs and looking for signs of smoke or fire.

Wally Crippen collection
Port Angeles and Western Railroad hauled spruce logs on trains like this in the 1940s.

The financial difficulties apparently contributed to a second set of problems for the railroad. Its infrastructure began to deteriorate. Portions of the railroad tracks fell into disrepair. Grass, weeds and other vegetation grew between the tracks' ties, along the tracks and on the railroad right-of-way. Discarded, rotten ties accumulated along the right-of-way.

The condition of the railroad and its questionable fire prevention procedures exacerbated the risk of wildfire at a time when fire conditions were already extreme.

Then, shortly before noon on August 6, 1951, a steam-powered train hauling 18 cars of logs and freight headed east through the Olympic forest. With the crew apparently unaware, small fires flared up along the right-of-way, in all likelihood ignited by sparks or heated metal from the locomotive. The train pulled alongside a water tank near Camp Creek to take on water and to give the four-man crew time to eat lunch. About 20 minutes after the train stopped, crewmen spotted a fire under one of the cars. The crew backed up the train, grabbed fire tools and water and doused the flames.

The train continued to head east, and the crew noticed more smoke. Pulling into a stop, the conductor reported the smoke to the U.S. Forest Service at the Snider Ranger Station northeast of Forks. By then, a forest fire lookout had already reported the smoke to District Ranger Sanford M. "Sandy" Floe, a 54-year-old veteran at the Snider Ranger Station. Floe, responding to sightings of several fires along and near the railroad tracks, coordinated a fire-fighting effort that involved crews from the U.S. Forest

The Great Forks Fire

Service, the Washington State Division of Forestry, Rayonier, Fibreboard and the railroad.

Forks Timber Museum

The August blaze consumed 1,600 acres, then erupted again in September.

Still, the fire spread. The following morning more firefighters were called to the scene; Fibreboard alone sent 45 to 50 men and bulldozers. Firefighters tried to contain the blaze to 60 acres, but the fire was out of control by mid-afternoon. The runaway blaze jumped across fire lines and burned about 1,600 acres of forestland, including property owned by Fibreboard.

The blaze was finally brought under control on the afternoon of August 10, although spot fires persisted. Fire crews spent the following weeks mopping up the blaze. By mid-September, the Forest Service reported no visible signs of smoke.

But some of the fire lived on, perhaps smoldering, undetected, just below the forest floor, or in a stump, or a fallen log. Perhaps a standing snag harbored the stubborn coals. Nobody knows for sure, except that the blaze was never completely extinguished.

Then, on the evening of September 19, the weather changed.

Humidity dropped, and a strong wind of "unusual and extraordinary intensity" started blowing out of the northeast. More often, prevailing winds on the Olympic Peninsula come from the opposite direction (from the southwest, off of the Pacific Ocean), and the ocean-driven weather systems are often sloppy with clouds and rain.

On that night, however, a dry wind out of the east created a dangerous situation. Somehow, the wind delivered fresh oxygen to the remaining coals, which glowed more brightly and burst into flames. The easterly wind became the equivalent of a blacksmith's bellows.

By daybreak of September 20, the smoldering fire, fueled by great amounts of old and new logging debris, turned into a rapidly spreading blaze that raced through the

The Great Forks Fire

trees and pumped great clouds of reddish-brown smoke into the morning sky. Powered by what was described as "gale-force" winds, the fire headed down the Calawah Valley toward Forks.

Glen S. "Mickey" Merchant was working the overnight shift as a fire lookout at the State Division of Forestry post on Gunderson Mountain, located west of Highway 101 and roughly four miles north of Forks. As a lookout, Merchant had the job of watching for fires from a lookout cabin perched on the mountain. Merchant was an expert woodsman, trapper, hunter and legendary fly fisherman; reportedly, he once fought off a cougar with a fishing pole.

That night at about 3 a.m., Merchant surveyed the scene from his observation cabin. He looked across the Sol Duc River Valley to the Calawah Ridge and beyond that to the foothills of the Olympic Mountains. Roughly 18 miles from Forks, the horizon was glowing red.

A better view of the fire was available at the North Point Lookout Station, a 12-year-old U.S. Forest Service observation post about 20 miles northeast of Forks. But on that night, of all nights, the North Point Lookout was empty. So Merchant reported the fire to John LeRoy

"Mac" MacDonald, the district warden for the state forestry division.

Glen S. "Mickey" Merchant as a young man.

MacDonald, then in his mid-50s, was an early riser with lots of contacts in the timber industry. He immediately notified the state and the timber companies – most likely Merrill & Ring, Bloedel-Donovan, Crown Zellerbach and Rayonier – about the fire. Then he got in his truck and headed out Highway 101, rounding up his crew, recruiting loggers and borrowing bulldozers.

At about the same time, the U.S. Forest Service responded too. Edward G. "Ed" Drake was awakened by the sounds of somebody banging on the door of his bunkhouse at the Snider Ranger Station northeast of Forks. Drake's boss wanted him to check out reports of a fire.

Drake, a 23-year-old forestry student at the University of Washington and seasonal forestry worker in the Olympics, got in a Forest Service pickup and headed east toward the burned area where the fire had erupted in August. Before Drake even got out of the truck, he saw the flames, which he likened to an enormous campfire that lit up the sky. "I could read a newspaper just by the light of the fire," he recalled.

He returned to the ranger station to report the blaze, but District Ranger Sanford Floe was already telling his superiors at the Forest Service's regional headquarters in Portland, Oregon, that the fire was a big one. "Send everything you've got." Floe said. That meant prepare for the worst, with shovels, axes, sleeping bags, tents – the kind of equipment necessary to put down a large forest fire.

Ranger Floe and his 11-year-old son, Sandy, got into a heavy-duty pickup truck to help evacuate a nearby family whose ranch was threatened by the fire. Young Floe watched as trees, pelted by embers, went up in flames. It was hot, and the wind was howling. The wind was so strong, he recalled, it blew empty 55-gallon oil drums down the road.

By 6 a.m., Ranger Floe and his son stood on a road leading to a lookout above the Snider Ranger Station. There the two had a sweeping view of the Sol Duc and Calawah valleys. In later years, Ranger Floe would recall watching the scene that morning as reportedly the "saddest point" in his long career.

The easterly wind was fanning the fire and driving the flames through the forest, which was filled with large cedar, Douglas fir, hemlock and spruce.

The blaze was moving so fast that Floe's fire crews didn't have time to establish a line of defense. "It blew over the top of our heads and jumped ahead of us," recalled Llewellyn J. "Lew" Evans, Floe's assistant ranger.

Some firefighters moved in behind the fire, but their efforts were futile, and they were forced out. Because it was a wind-driven blaze, Floe didn't want to risk his firefighters' lives by ordering them to set up positions in front of it.

Floe watched, helpless, as the fire raced past the Snider Ranger Station and roared toward Forks, about 18 miles away.

CHAPTER 3

Frontier Spirit

Twenty-year-old Lawrence Gaydeski was logging hemlock and spruce trees early that morning at a site near the Hoh River, about 12 miles south of Forks. His crew, with two or three Caterpillar tractors, or "Cats," had already been hard at work for several hours on their "hoot owl" shift.

About 8 a.m., Gaydeski's boss pulled up to the worksite and alerted the crew to the fire several miles away, in the Calawah Valley. The boss shrugged off rumors that Forks residents were evacuating town. "I don't think there's anything to worry about," he reassured the crew.

An hour later, all nonchalance vanished. Logger Bill Howard drove up in a flatbed truck and yelled: "Get your Cats and bring 'em to Forks." The plan was to bulldoze ditches or furrows on the outskirts of town, and make fire lines to keep the fire from advancing into Forks. "We're moving every Cat we can find in the forest," Howard said, and with that, he confiscated a bulldozer.

The Great Forks Fire

It was not unusual for people in the Forks area to take charge. The frontier spirit of self-reliance and individualism was part of the makeup of Forks residents in the 1950s. They were used to being responsible for themselves and their families. They believed they could take care of their own.

Pioneers and homesteaders who settled the Forks area had to be tough. They carved out a living in a densely forested, thinly populated corner of the Northwest. Isolated and often lonely, they lived with interminable rain and the threat of bears, cougars, wolves and wildcats. Basic urban services were distant luxuries. Gaydeski recalled one 50-year-old man who pulled his own tooth with a pair of pliers. "He was a tough old Swede," Gaydeski recalled.

People still talked about the legendary John Huelsdonk, the "Iron Man of the Hoh," a pioneer who lived south of Forks in the Hoh River rainforest. The unusually strong Huelsdonk was reputed to have hiked a wilderness trail to his homestead one day with a cast-iron stove on his back. In an encounter that is now the stuff of legend, a forest ranger met Huelsdonk and commented on his heavy load. As the story goes, Huelsdonk said he didn't mind the stove – it was the 50-pound sack inside that gave him trouble.

Lawrence and Dixie Gaydeski with their son, Gary, in a 1950s photograph.

The same frontier spirit and self-reliance were evident on the day of the Great Forks Fire. Without bosses barking orders, men took it upon themselves to haul in Cats and bulldoze fire lines to try to control the flames.

The fire was on a clear path to Forks. As the blaze advanced, crews brought in flatbed trucks to move families and belongings to safe ground. Loggers, more at home driving a logging truck or wielding chainsaws or cross-cut saws, loaded trucks and cars with baby cribs, appliances and furniture. They saw a need and responded. People helped one another.

For Gaydeski, it was a time to pitch in. After hurrying home from the Hoh River and seeing to the safety of his wife, Dixie, and son, Gary, Gaydeski drove to Forks to help his mother, Louise, who ran the Evergreen Cafe in town. She recently had taken delivery of a large supply of groceries for the winter. With the fire coming dangerously close, she needed a safe place to store her groceries.

Strong winds pushed the smoke ahead of the blaze, and smoke filled the air as Gaydeski loaded his vehicle with his mother's restaurant supplies. He drove the supplies to his home on the Sol Duc River, unloaded, returned to Forks, loaded the vehicle again and made another trip. Later, he worked the fire lines.

Log-truck driver Jim Mansfield rushed home from a Hoh River worksite and dug a trench to keep the fire off his property. The 21-year-old Mansfield also helped the state forestry division pump precious water from a creek alongside his property into a state water tanker for firefighters to use.

Mansfield's wife, Pat, helped strip beds at the Forks Hospital so that patients could be moved to Aberdeen, more than 100 miles south of Forks. Pat's mother fixed sandwiches and made coffee for the fire crews. Volunteers hitched rides on the local ambulance and brought coffee and water to men building a defensive fire line at the north end of town.

"The way we were all brought up in those days, you either did it yourself or it didn't get done," recalled Jim Mason, who was a 21-year-old log-truck driver at the time of the Great Forks Fire. "People, particularly in the smaller areas, want to help each other."

But the blaze continued to advance toward the town. By late morning, the sky over Forks was dark. Thick smoke blocked out the sun. In the darkness, the streetlights went on.

CHAPTER 4

The Fire

Forests are fuel to fires. From the twigs on the forest floor to the towering Douglas firs, the forests of Western Washington are composed of organic material that burns under the right conditions. Even a log as big as a four-door sedan can be reduced to a line of ash.

Logging sometimes leaves more fuel on which a fire can feed. Loggers in the 1950s, in comparison with their counterparts in later years, left behind more discarded tree limbs and other woody debris, or "slash," which can get tinder dry over the course of a summer. Often, too, logging companies left behind stacks of logs destined for markets.

The other chief ingredient of fire is oxygen, and wind can feed the flames like a bellows. A strong wind pours oxygen into the fire and pushes the flames toward fresh fuel. Wind also blows the heat ahead of a fire, drying out the trees and the underbrush, making the forest even more vulnerable to the advancing flames.

Forest fires even create their own wind. The hot air

rises into the sky, creating a powerful updraft. It works like an atmospheric chimney, sucking in fresh air from the ground and heating it in the furnace of the fire. The superheated air boils toward the heavens in a dramatic thermal column.

Wally Crippen collection

Loggers in earlier years left behind more flammable "slash," or scattered wood debris, as seen in this 1950s Hoh River logging scene.

Columns filled with hot air and smoke can soar as high as 20,000 feet, dwarfing even the highest peaks of the Olympics and the Cascades. On the ground, the updraft can be terrifying and surreal. A major fire will create a vacuum so strong that it can suck aloft six-foot logs and small boulders in a swirling fury that some liken to a tornado.

The longer a fire burns, the faster it moves, and a strong wind can whip the maelstrom forward at even greater speeds. The fire's powerful updraft strips the forest of burning debris and hurls the embers thousands of feet into the air. Once aloft, the embers can ride a strong wind a quarter mile or more. They eventually fall to earth ahead of the main blaze, often creating numerous "spot fires."

Animals sense the danger, and they flee. But they do not necessarily know where to run, and the animals often get trapped when rugged terrain or spot fires block their escape from the main fire. Many die.

The noise can be deafening. The sound of a forest on fire, and the turbulence it generates, has been compared to the clatter and roar of a passing freight train. Experienced firefighters say they can hear a forest fire coming.

The east wind played a powerful role in the Great Forks Fire. The easterly was very dry and often punishing, roaring across the treetops as fast as 30 to 40 mph. And on the day of the fire, flames marched ahead of the wind, leaping from treetop to treetop, a phenomenon known as "crowning." Some reports said the flames leapfrogged a mile at a time.

Flames shot up in the air, some 400 to 500 feet. One witness described old Douglas fir trees, laden with pitch, becoming "huge torches of reddish-black flames and smoke." Another recalled that it was a "firestorm."

The strong winds flung glowing embers a half mile ahead of the main blaze. Burnt needles, small branches and chunks of burning bark, carried aloft by the powerful winds, pelted rooftops and yards in Forks throughout the day.

As the flames advanced toward Forks, more residents retreated to safety. Hundreds of people piled belongings into their cars and pickups and left in an exodus that reminded some observers of a scene from the movie "The Grapes of Wrath."

The main road was Highway 101, which goes from Port Angeles through Forks and on to Aberdeen. The northern route out of Forks took travelers north, then to points east – to Beaver, Sappho, the Snider Ranger Station and Port Angeles. But the fire was coming from that direction, making the route more perilous. The southern route went past the Bogachiel and Hoh rivers and on to Aberdeen. But even that route was hazardous when sparks started flying over the Bogachiel.

By lunchtime, Ella Brager, her daughters and their dog and cat were on Highway 101 heading south in a station wagon to Ella's mother's home near Yelm, east of Olympia. Ella's daughter Nancy, thinking she'd never see her father again, started crying. "He's got a walkie-talkie," Ella soothed her daughter. "He can go to the river and protect himself."

Smoke obscured the sky, and an eerie, reddish twilight descended over the town. "We didn't know what we left behind" in Forks, Ella recalled. "I didn't think I'd be back. I thought the town would be gone."

Like the Bragers, many families used Highway 101 and drove south to Aberdeen. Others went to the Quillayute Naval Air Station west of town. Early evacuees escaped to Port Angeles before smoke and falling embers made the highway too perilous. Flames in the foothills were visible along Highway 101.

Some motorists who braved the 58-mile drive to Port Angeles saw smoke from the blaze that soared thousands of feet into the sky and blotted out the sun. Some said the smoke billowed 15,000 feet above the blaze. The smoke was so thick it blackened the sky. Motorists peered through their windshields into the gloom and turned on their headlights.

The Great Forks Fire

Relentlessly, the strong winds continued to drive the fire. The fast-moving blaze raced at a tremendous speed. Reports vary, but the fire traveled about 18 miles in less than a day. Flames tore into giant trees. Ash particles fell from the sky.

The Seattle Times

A lone person stands on a stump at the outskirts of Forks and watches as the fire roars closer to town.

Forest Service fire control crews worked at the edges of the blaze to try to keep the fire from spreading sideways. But there was little the men could do to keep the fire from advancing to Forks.

Forks was a town under siege. The fire was buffeting the town on three sides – east, west and north. Especially vulnerable was the north end of Forks, which was directly in the path of the oncoming fire. Worse, the area was dotted with shrubs and vegetation – fuel for the sparks and embers that were flung by the blaze.

Ted Spoelstra and his brother John had a truck and equipment shop at the northern edge of the business district. Worried, the Spoelstra brothers loaded as much as they could from their shop and drove their equipment to safety, south of town. "We had to leave the location at 11 a.m.," Ted Spoelstra recalled. "It was so hot and smoky, you couldn't see."

By noon, the fire was lapping at the outskirts of town.

CHAPTER 4

Forks Fights Back

Throughout the day, ordinary people with no firefighting experience – shopkeepers, homemakers, retirees – were confronted by the advancing inferno. Many made heart-wrenching decisions about their personal safety and property, with little time to ponder the consequences. Others simply wanted to save their town. Over and over that day, ordinary people took enormous risks.

As the fire raged closer, people from all walks of life built fire lines to divert the fire. They hosed down buildings and poured water on falling embers. They wetted gunnysacks and laid them on roofs. They patrolled streets and stamped out sparks and embers. They swept up burnt fir needles. They worked side by side with firefighters from Forks and other Peninsula towns. The day called for heroism, and Forks answered the call.

They took care of their own.

Russ and Helen Thomas loaded a logging truck with their most prized possessions, including their beloved

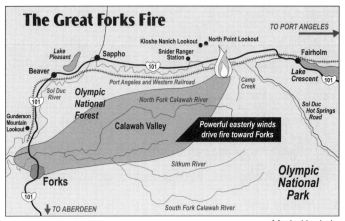

Marin Henkels

washing machine. Once his family and belongings were safely on their way, Russ stayed behind and protected his home and business. "I told a patrolman, 'I'm going to protect my property,'" Russ recalled. "I'm going to stay here."

He got a broom and swept clean the roofs on his business and home, which were across the street from each other in downtown Forks. Then he got a garden hose and poured water on the rooftops as the fire closed in.

A news report tells of a young man who gathered up his family, including his aging father, and took them to a safe place. When the young man turned to head back to town and fight the fire, his elderly father, despite having heart trouble, insisted on going too. "Dad, remember your heart," the young man cautioned. "My heart is in Forks," the older man replied. "I am going with you. Let's get rolling." Later that day, the father and son helped save two homes from the flames.

At the Kaiser-Frasier auto dealership in Forks, owner Art Anderson realized the flames were bearing down on his car lot. Some cars were so new that Anderson hadn't had time to remove the protective paper.

At the same time, people without cars were desperate to get out of town. Anderson gathered up the keys to the cars, gassed them up and loaned the vehicles to anyone who needed a car to evacuate.

Anderson watched his dealership's cars roll down Highway 101 one by one and disappear out of sight.

By 2 p.m., fire officials believed the advancing flames would consume the town. They called for an immediate evacuation, fearing that residents would either suffocate from the smoke or die in the flames if they stayed. State Highway Patrol Sgt. Clifford Aden ordered everyone, except those fighting the fire, to leave Forks. Highway 101 between Forks and Port Angeles would be impassable within minutes, Aden said. A truck equipped with a loudspeaker blared warnings for people to leave immediately.

The Great Forks Fire

News reports later estimated that 85 percent of Forks' 1,100 residents fled the town.

Among the last to leave were Edwin and Peggy Beebe, childhood sweethearts who had married three years earlier. They had a daughter, two-year-old Darla Jean, and another child was on the way.

The young couple loaded everything they could into their dump truck and headed northeast on Highway 101, toward Beaver. By now, flames were at the side of the road. But the Beebes pressed on.

"We were very much in love," the young wife and mother recalled. "We were just concerned about getting everything out we could at the time, getting our girl out safely."

Earl Clark was in Forks that day reporting on the biggest story of his career as city editor of the *Port Angeles Evening News,* now the *Peninsula Daily News.* "The immensity of it didn't shape up until we got to town," Clark recalled. "We saw a wall of fire encroaching on the town."

Clark later filed a story for the newspaper. Describing the scene, Clark wrote, "Sparks and embers rained through the air, lighting on shingle roofs, in dead snags, in piles of kindling. Fanned by a strong east wind, which the heat of the fire generated into small whirlwinds, the embers set off hundreds of fires everywhere they lit."

Thirty-eight-year-old Vic Ulin was chief of the Forks volunteer fire district, which was woefully unequipped for such a formidable blaze. The community had a couple of fire trucks and a water system that could muster only 10 pounds of pressure under the best of circumstances.

Residents pushed the water system to the limit by pouring water on their roofs. Ulin patrolled the town in one of the district's fire trucks. "There was nobody sitting still anywhere," Ulin recalled.

Even as residents fled, the two restaurants in town teamed up to fix sandwiches and coffee for the firefighters throughout the day. And the weary men, their bodies stoked by coffee and determination, kept going.

Bulldozer crews and men wielding shovels swarmed to the edge of town in an effort to hold back the fire by building a defensive line. Firefighters from Port Angeles, Aberdeen and other towns manned pump wagons. Fire crews, armed with water hoses, pumped water from the Calawah River and doused homes and buildings. Townspeople worked feverishly alongside the firefighters to help quell or divert the fire. Though their efforts helped prevent spot fires from flaring up, nothing could stop the main blaze from advancing.

"There were only three kinds of sounds," Clark wrote, "the roar and crackle of the flames eating into the town, the scream of sirens as fire trucks raced through the empty street, and the urgent shouts of men's voices."

The blaze closed in.

At about 2:15 p.m., the fire tore into the edge of town. Then, at 2:30 p.m., the fire jumped across the Calawah River north of town and leaped across Highway 101. It was a calamitous event. The fire had crossed the highway, cutting off the northern escape route to Port Angeles.

Flames now separated 23-year-old Ron Shearer from his family. Earlier that day, Shearer and his brother, Reed, had taken their parents and Shearer's wife, Marjorie, to safety in Sappho, about 13 miles northeast of Forks. The brothers then returned to Forks to fight the fire, ultimately to see the blaze cut them off from their loved ones.

"It was hard leaving my wife and family, my mother and dad," recalled Shearer. "You feel responsible. Then when you find out the road's closed, you worry. My wife couldn't drive at that time. You look up the valley. There's a tremendous fire in the valley."

Still, Shearer and his brother joined with firefighters, loggers, mill workers and other townspeople in a common goal, he said, "to save our town, to save our home, our hospital, our way of life."

But reminders were all around them that this blaze was bigger than anything the men could throw at it. Lifted by the fire's turbulence, chunks of moss-covered bark as big as dinner plates floated through the air, recalled Shearer.

"On the east edge of town, fire leaped from fiercely burning trees into adjacent houses and barns," city editor Clark wrote. "A couple of fire trucks with booster tanks stood in the yards while firemen braved the intense heat," and aimed their "puny streams of water on the blaze."

The heat from the fire was "roasting the town." Working in the heat was grueling for the men fighting the fire, but they kept at it, doing whatever they could. Some of the men had been rousted from their jobs early that day, and their faces were now streaked with dust and grime. But they persevered. They took care of their own.

By now, however, some people in town were losing hope, saying they wouldn't give "two bits" for Forks.

The Great Forks Fire

It looked like the fire, driven by the relentless wind, would destroy the town. One of the first buildings to go up in flames was a barn at the outskirts of Forks. Not long after, the adjacent farmhouse burned too. Advancing flames consumed tourist cabins, a house, a garage, a sawmill, a supply of cedar logs – even the Spoelstra brothers' shop, which they had evacuated only that morning.

"It looked like a doomed town, because the fire was burning up houses right in front of our eyes," city editor Clark later recalled. At the time, he wrote: "In the north end of town, the flames exploded houses like matches . . . Firemen dashed around the roofs extinguishing embers, which rained down like hail."

By 4:30 p.m. the only escape route was to the south toward Aberdeen.

Sometime later, MacDonald, the district warden, left his post near Beaver and drove on Highway 101 to Forks, exhausted. His back ached.

As MacDonald approached the edge of Forks, he was horrified, recalled Marjorie Jensen, his daughter. Thick smoke enveloped the town, which was directly in the path of the advancing flames. Homes and businesses were burning on the outskirts of Forks.

The fire and the easterly wind were bigger than the men and their machines. Without a miracle, Forks was doomed.

MacDonald looked over the scene. His heart sank.

"Dear God," he said. "Help us."

Memory is a supple thing, and often unreliable, especially from the distance of 50 years. But many people who were in town that day share the same memory of what happened next.

The hot, relentless wind from the east gradually eased, they say. Then, out of the west, came a fresh wind that blew in off of the Pacific Ocean, cool and moist, and just strong enough to stop the fire's advance.

The town was spared.

CHAPTER 6

Epilogue

By nightfall, men patrolled the streets of Forks and continued to build fire lines. Though the worst was over, the fire threat remained. One news report described the scene as a nightmare: "Fire-fighters roved through a red, smoky pall like ghostly figures in the deserted town. Flying [embers] ignited sometimes a half dozen blazes at once."

Mercifully, the path of the fire had shifted, and the wind now drove the fire to the southwest, toward Mill Creek and the cool waters of the Bogachiel River south of Forks. "If the wind hadn't shifted, swinging the force of the fire away from the town itself, no human effort could have saved Forks," wrote city editor Clark. "It was a miracle of nature, aided by the dogged determination of weary firefighters who just wouldn't give up, that saved Forks from total destruction."

The next day, MacDonald, the district warden, and Floe, the district ranger, divided up tasks for their fire crews, roughly 450 men. Some were sent to bulldoze fire

The Great Forks Fire

roads south of Forks. Other firefighters used chainsaws to cut down smoldering snags, or dead, standing trees. Still others went after spot fires.

Clallam County Historical Society
Smoke hovers over a scene of the fire's destruction.

Clallam County Historical Society
A burnt car is among the wreckage of the fire.

Seventy-one-year-old Oliver J. "Ollie" Ford protected his homestead with a garden hose. Ford, whose parents pioneered the Forks Prairie in 1878, sat on the porch of his home, which was in sight of the log cabin where he was born and barely 200 yards from where flames had struck the day before. Ford held a garden hose in his lap all day and dared the fire to come his way.

A few days later, gentle rain fell in outlying areas and the foothills. Before long, the fire-scarred town experienced a welcome sight. The dry days were over, and the seasonal rain returned to Forks.

Remarkably, no one was killed or seriously injured in the Great Forks Fire, despite hundreds of men and women working in dangerous and difficult situations.

The financial toll, however, was considerable. The fire charred about 33,000 acres. The U.S. Forest Service estimated that the blaze damaged federal timber that was worth about $1.5 million in 1951 values and dollars. The burned trees contained about 600 million board feet of lumber, an amount worth approximately $5 million at the time. Fighting the fire wasn't cheap either; the Forest Service figured its share of the bill came to $140,000.

The fire also blackened state and private forests. Rayonier lost a bridge, and suffered damages to its logging and railway equipment, according to court documents. The company also suffered losses to its stands of timber.

Hundreds of acres of old-growth were in the path of the fire. The Gourlay Lumber Co. lost its sawmill, valued at $55,000. All of the dollar amounts are stated in 1951 values.

The most serious damage was at the north end of town, where Highway 101 enters Forks from Port Angeles. News accounts vary, but roughly two dozen families were completely wiped out. Their losses included their homes, furniture and clothes. Others lost barns, garages, trucks, tools, heavy equipment and logs in the blaze. Losses were estimated at more than $1 million, a hefty sum at the time.

Some families were particularly devastated. Jim Mason, a log-truck driver, his wife, Helen, and infant daughter, Terry Lin, had moved into a new home only three months before. "We wanted to buy it," Mason said. "We started putting all our money into furniture. We lost all of it – an automatic washer, refrigerator, dressers, beds" and living room furniture.

Ted Spoelstra and his brother John were able to save tires, spare parts and other items from their truck and equipment shop before the fire consumed it, but the brothers left behind a lathe and other equipment, which went up in flames. It was an extensive loss, Ted Spoelstra recalled, but he and his brother were young and resilient. "We rebuilt the place," he said.

Not far from the Spoelstras' shop, John and Rue Maxwell operated a nine-unit tourist court and lived in an adjoining home. The couple wet down their home and saved it, but their tourist cabins were destroyed – a loss of about $12,000 in property and furnishings, according to court documents.

The Seattle Times

John and Rue Maxwell hose down the charred ruins of their tourist cabins.

Others fared better. Russ Thomas, armed with a garden hose, saved his home and business.

Edwin and Peggy Beebe and their daughter, Darla Jean, returned to town worried about their home and belongings, their truck still piled high with furniture. Their home was still standing. "We were blessed," Peggy Beebe said.

At her mother's house near Yelm, Ella Brager got a telephone call from her brother-in-law. "You still have a home," he said. But the Bragers lost logging equipment and a stack of logs ready for the market.

After the smoke cleared and the danger subsided, residents who fled the town returned. One by one, residents who had borrowed a car from Art Anderson drove the vehicle to the dealership, parked it and returned the keys.

He got back every single car.

In the fire's aftermath, many paid tribute to the people in Forks who wouldn't give up. Their heroic work proved to be the tipping point – the margin of time the town needed to allow nature to run its course and shift gears. Hosing down roofs, digging trenches, and stamping out embers made a difference, and most likely prevented more spot fires from igniting and spreading.

"Give a lot of credit to those great firefighters who saved the town and to the communities that sent fire fighting equipment and men to help us," said Dr. Ulric S. Ford, mayor of Forks.

"The fact that the town is still there – most of it at least – is a tribute to the dogged persistence and guts of a couple of hundred sturdy men," wrote city editor Clark.

"We couldn't have saved the town without them," said Oscar Herd, acting fire chief of Forks.

They took care of their own.

****

In December 1951, about 20 households and businesses in Forks sued the U.S. Forest Service, Fibreboard Products Inc., and the Port Angeles and Western Railroad Co. for their losses in the September blaze.

The lawsuit, *Arnhold et al. v. United States of America and Port Angeles & Western Railroad,* claimed that the fires on August 6–10, 1951, which led to the Great Forks Fire, were preventable and should have been fought more aggressively.

Shortly afterwards, the railroad filed for bankruptcy protection. The case dragged on for years, but the residents of Forks who sued eventually prevailed when the Forest Service was held liable for damages in U.S. District Court in Seattle.

The Great Forks Fire damaged a huge swath of public and private forest filled with trees that still had some commercial value, provided that they were logged and hauled to mills expeditiously.

Foresters estimated that they could salvage more than 600 million board feet of timber – an enormous volume for the area – from the blackened public and private forests. The U.S. Forest Service in January 1952 announced plans to sell 425 million board feet of damaged federal timber within two years, exceeding, by far, its sustained-yield logging schedule of 51 million board feet a year.

Fire-damaged timber flooded the market. By 1955, most of the timber had been sold to buyers at discount prices and hauled away to the mills.

As the damaged trees were removed, landowners began replanting the forest. But they changed the character of the landscape by planting only Douglas fir, and the sheer scale of the reforestation forced them to bring in seeds from outside the area. As a result, the forest that grew after the fire was different, with less diversity and more non-native trees. Decades later, the forest is still different, another legacy of the Great Forks Fire of 1951.

Acknowledgments

I want to gratefully acknowledge the help and support of the following people; George Erb, Marin Henkels, Veronica Weikel, Roberta Sobotka, Tom Thompson, Lonnie Archibald, Phillip Mohammed, John Brewer, Stan and Sherrill Fouts, Dave Cole, Wally Crippen and Bill and Susan Brager.

I am especially grateful to the many people who shared with me their first-hand accounts of the Great Forks Fire, including Sandy Floe who, as an 11-year-old boy, accompanied his father, U.S. Forest District Ranger Sanford M. "Sandy" Floe, in the early hours of the fire; Edward G. "Ed" Drake, the young Forest Service seasonal worker who hurried from Snider Ranger Station to confirm the fire; and Earl Clark, the city editor of the *Port Angeles Evening News,* who was in Forks that eventful day.

Many of the people who helped tell this story still live in Forks. Ella Brager, the young housewife who fled the fire with her daughters, was widowed and the mother of three when she married Warren Paul in 1962. He died in 2003.

The Great Forks Fire

Warren and Ella Brager Paul

Peggy and Edwin Beebe

Ted Spoelstra

Ron and Marjorie Shearer

Lonnie Archibald photographs

Edwin and Peggy Beebe, who live in the same house from which they fled on the day of the Great Forks Fire, divide their time between Forks and Arizona.

Ted Spoelstra, whose truck and equipment shop was destroyed in the blaze, served as commissioner for the Port of Port Angeles from 1985 to 1991.

Lawrence Gaydeski, the logger who rushed home from the Hoh River to see to the safety of his family, still lives on the Sol Duc River with his wife, Dixie. Gaydeski served as Clallam County Commissioner from 1983 to 1994.

The Great Forks Fire

Jim and Pat Mansfield

Vic and Helen Ulin

Helen and Russ Thomas

Dixie and Lawrence Gaydeski

Russ Thomas, who protected his home and business from the fire, lives with his wife, Helen, in a ranch on the edge of town. Vic Ulin, the fire chief who patrolled the threatened community, has lived in the same house in Forks for more than 60 years. Ulin lost his wife, Helen, in 2001.

Also, still living in Forks are Ron Shearer, who was separated by flames from his wife, Marjorie; and Jim and Pat Mansfield, who helped state forestry officials pump water and assisted in the evacuation of Forks Hospital.

Art Anderson, who loaned cars to families evacuating the town, continued to run his automotive business after the fire.

Notes

Chapter 1, "It's a Fire"

Ella Brager was up early; Ella Brager Paul, interviews by author, August 10–11, 2000, January 26, 2001, September 19, 2002.

Chapter 2, A Dry Summer

Sources for this chapter were drawn almost entirely (exceptions are noted) from the civil case files *Arnhold et al. v. United States of America and Port Angeles & Western Railroad*, and *Rayonier Inc. v. United States of America*.

U.S. forestry officials; *Arnhold* and *Rayonier;* See especially Docket #152; also #146.

Just two years before; For a true account of the Mann Gulch Fire, see *Young Men and Fire*, by Norman Maclean.

independent loggers, called; Gyppos is also spelled gypos.

The railroad had lost; The deterioration of the railroad was cited numerous times in *Arnhold*. See especially Dockets #57 and #146. See also Harry LeGear's account of the railroad in the *Jimmy Come Lately history of Clallam County,* pages 83–91. See also Paul Martin, *Port Angeles, Washington: A History,* pages 129–131.

Financial difficulties; *Arnhold,* also see LeGear.

steam-powered train; *Arnhold,* Docket #57, also #146. The train was Engine No. 1347.

sparks or heated metal; *Arnhold,* Unnumbered document, U.S. Government/U.S. Forest Service. In correspondence to the author December 3, 2002, Ted Spoelstra suggests the fire may have been ignited by a "hot box" or an overheated bearing on one of the logging cars. In a *Port Angeles Evening News* report September 24, 1951, Olympic National Forest Supervisor Carl B. Neal attributed the sparks to a "loose pin" on the locomotive "rubbing against a wheel flange."

a forest fire lookout; *Arnhold,* Docket #146.

The Great Forks Fire

"unusual and extraordinary intensity"; *Arnhold,* Docket #20.
fueled by great amounts of; Retired U.S. Forest Service civil engineer Stan Fouts, correspondence to author, November 2, 2002.
"gale-force" winds"; *Arnhold,* Docket #152.
"Glen S. "Mickey" Merchant"; Merchant's deposition is missing from court documents; however, it is possible to infer his actions the morning of September 20, 1951, from a list of questions in Docket #109 *Arnhold,* and from contemporary newspaper accounts. Personal recollections are from interviews with his niece Daisy Sinnema and with Lawrence Gaydeski.
So Merchant reported the fire; *Arnhold,* Docket #146.
At about the same time; Edward G. "Ed" Drake, interview by author, January 15, 2003.
Ranger Floe and his; Sandy Floe, interview by author, January 2, 2003.
"saddest point"; Ranger Sanford M. "Sandy" Floe, quoted in Jack Rooney's *Frontier Legacy,* page 86. See also C.N. Webster's column, *Port Angeles Evening News,* September 21, 1951.
"It blew over the top"; Retired U.S. Forest Service District Assistant Llewellyn J. "Lew" Evans, interview by author, January 15, 2003.
Some firefighters moved; Fouts, correspondence to author, November 2, 2002. Fouts recalled how his father, Creighton "Slim" Fouts, and two other firefighters, tried to save three large stacks of logs by pumping water from a creek behind the fire. Fouts said the men "spent most of their time eating smoke" through dampened handkerchiefs before giving up on their efforts.

Chapter 3, Frontier Spirit

Twenty-year-old Lawrence Gaydeski; Lawrence Gaydeski, interviews by author, August 22, 2000, March 15, 2001, December 14, 2002. Also see videotape *Out of the Ashes,* the 2nd annual Old Timers' Roundtable, Forks Heritage Days, October 2–10, 1998.
he confiscated a bulldozer; Gaydeski. Also Ted Spoelstra,

interviews by author, September 11, 2000, December 14, 2002, January 7, 2003. Spoelstra credits Frank Henry and Bill Howard with saving the town by taking the "bull by the horns" and bringing in the Cats. After the fire, critics complained that the fire damage would have been less severe if State Fire District Warden John LeRoy "Mac" MacDonald had allowed people to bring in more tractors. But MacDonald defended his actions in an interview with the *Port Angeles Evening News* that appeared October 17, 1951. "We didn't stop anyone from bringing in a bulldozer who wanted to," MacDonald said. "But after we had 10, we didn't hire any more. They did clear a fire trail, until it got too hot to work." MacDonald said it would have endangered lives and equipment to "clear any more area than we did."

the legendary John Huelsdonk; See Elizabeth Huelsdonk Fletcher's account in *The Iron Man of the Hoh,* page 152. Also see Murray Morgan's *The Last Wilderness,* page 154; and *Footprints in the Olympics, an Autobiography* by Chris Morgenroth, page 52.

50-pound sack; Some say it was 100 pounds; also some versions say it was a sack of sugar, others say it was flour.

Later, he worked the fire lines; Gaydeski.

Log-truck driver; Jim Mansfield and Pat Blevins Mansfield, interview by author, August 22, 2000.

The way we were all brought up; Jim and Helen Mason, interview by author, August 28, 2002.

Chapter 4, The Fire

Sources for this chapter were drawn in large part from interviews conducted August 10, 2000, with Jack Zaccardo, retired forester, and David Cole, forester, Washington State Department of Natural Resources (formerly Washington State Division of Forestry). Also David Cole interviews August 11, 2000 and February 17, 2001.

400 to 500 feet; Ron Shearer, interview by author, September 21, 2002.

"huge torches of reddish black flames and smoke"; Fouts, personal correspondence, April 7, 1999.

It was a "firestorm"; Ted Spoelstra.
By lunchtime; Ella Brager Paul.
Some said the smoke billowed; *Forks Forum,* September 27, 1951.
Reports vary; One account gives the fire's speed at 18 miles in six hours; another puts the speed at 14 miles in three and three-quarters hours. A *Seattle Daily Times* report on September 20, 1951, says the fire advanced 15 miles in seven hours.
Forest Service fire crews; Llewellyn J "Lew" Evans, interview by author, January 15, 2003.
"We had to leave the location"; Ted Spoelstra.

Chapter 5, Forks Fights Back

Russ and Helen Thomas; Russ Thomas, interview by author, August 23, 2000.
A news report tells of a young; Jack Henson column, *Port Angeles Evening News,* September 26, 1951.
At the Kaiser-Frasier auto dealership in Forks; Videotape, *Out of the Ashes.*
Among the last to leave; Edwin and Peggy Beebe, interview by author, August 11, 2000.
"The immensity of it"; Clark, interview by author, March 16, 2001.
"Sparks and embers rained"; Clark, *Port Angeles Evening News,* September 21, 1951.
Thirty-eight-year-old; Interview Vic and Helen Ulin, August 22, 2000.
"There were only three kinds"; Clark, *Port Angeles Evening News,* September 21, 1951.
Flames now separated; Interview Ron Shearer, September 21, 2002.
"On the east edge of town"; Clark, *Port Angeles Evening News,* September 21, 1951.
"roasting the town"; Forks Mayor Dr. Ulric S. Ford, quoted in *Seattle Post-Intelligencer,* September 21, 1951.

The Great Forks Fire

"**two bits**"; Quoted in Clark, *Port Angeles Evening News,* September 21, 1951.

"**It looked like a doomed town**"; Clark interview, March 16, 2001.

"**In the north end of town**"; Clark, *Port Angeles Evening News,* September 21, 1951.

By 4:30 p.m. the only escape route; *Seattle Post-Intelligencer,* September 21, 1951.

Sometime later, MacDonald; Marjorie MacDonald Cowan Jensen, interview by author, August 22, 2000.

Chapter 6, Epilogue

"**Fire-fighters roved**"; *The Seattle Daily Times,* September 21, 1951.

"**If the wind hadn't shifted**"; Clark, *Port Angeles Evening News,* September 20, 1952.

The next day; *Port Angeles Evening News,* September 22, 1951.

Seventy-one-year; Jack Henson column, *Port Angeles Evening News,* September 22, 1951.

The financial toll, however; *Arnhold* and *Rayonier.*

Rayonier lost a bridge; *Arnhold;* losses cited are according to court records, however news accounts estimate higher losses.

Hundreds of acres of old-growth; Spoelstra, correspondence to author, January 28, 2003.

The most serious damage; Clark, *Port Angeles Evening News,* September 20, 1952.

Some families were particularly; Mason interview. Quoted in *The Seattle Daily Times,* October 7, 1951. See also the videotape *Out of the Ashes.*

Ted Spoelstra and his brother; Spoelstra.

Not far from the; *Arnhold.*

Others fared better; Russ Thomas.

Edwin and Peggy Beebe; *The Seattle Daily Times,* September 23, 1951.

At her mother's house; Ella Brager Paul.
One by one; See videotape, *Out of the Ashes.*
"Give a lot of credit"; Quoted in Jack Henson column, *Port Angeles Evening News,* September, 22, 1951.
"The fact that the town"; Clark, *Port Angeles Evening News,* September 21, 1951.
"We couldn't have saved"; *Port Angeles Evening News,* September 21, 1951.
The lawsuit; Rayonier also filed suit, *Rayonier Inc. v. United States of America,* which was consolidated with *Arnhold* for trial. The case was heard by U.S. District Court Judge George H. Boldt who, in a controversial case in 1974, ruled in favor of tribal fishing rights.

Sources

Interviews by author

Edwin and Peggy Beebe, John Calhoun, Earl Clark, David Cole, Merritt Corbin, Wally Crippen, Edward G. "Ed" Drake, Keith Engelson, Llewellyn J. "Lew" Evans, Sandy Floe, Stan and Sherrill Fouts, Walt and Adria Fuhrman, Lawrence and Dixie Gaydeski, Marjorie MacDonald Cowan Jensen, Maynard and Pearl Lucken, David Mansfield, Jim and Pat Blevins Mansfield, Jim and Helen Mason, Jack Nattinger, Ella Brager Paul, Ron Shearer, Daisy Sinnema, Ted Spoelstra, Russ Thomas, Vic and Helen Ulin, Dick Welch, Jack Zaccardo.

Publications

Port Angeles Evening News, Seattle Post-Intelligencer, The Seattle Daily Times, Forks Forum.

Archibald, Lonnie, *There was a Day,* 1999.
Arnhold et al. v. United States of America and Port Angeles &

Western Railroad, and Rayonier Inc. v. United States of America.

Ficken, Robert E., *The Forested Land,* University of Washington Press, 1987.

Fletcher, Elizabeth Huelsdonk, *The Iron Man of the Hoh,* Creative Communications, 1979.

Kresek, Ray, *Fire Lookouts of the Northwest,* Ye Galleon Press, 1984.

Maclean, Norman, *Young Men and Fire,* University of Chicago Press, 1992.

Smith, LeRoy, *Pioneers of the Olympic Peninsula,* 1976.

Martin, Paul, *Port Angeles, Washington: A History, Volume 1,* 1983.

Morgan, Murray, *The Last Wilderness,* University of Washington Press, 1955.

Morgenroth, Chris, *Footprints in the Olympics, an Autobiography,* Ye Galleon Press, 1991.

Out of the Ashes, videotape sponsored by 2nd Annual Old Timers' Roundtable, Forks Heritage Days, October 2-10, 1998.

Rooney, J. R., *Frontier Legacy,* Northwest Interpretive Association in cooperation with the Olympic National Forest, 1997.

Russell, Jervis, ed., *Jimmy Come Lately history of Clallam County,* 1971.

Cover photograph: The Forks Timber Museum calls the photograph "The Fire that started the Forks Fire."

Index

Aden, Sgt. Clifford, 30
Anderson, Art, 30, 42, 47
Arnhold et al., 43

Beebe, Darla Jean, 31, 41
Beebe, Edwin, 31, 41, 46
Beebe, Peggy, 31, 41, 46
Bloedel-Donovan, 13
Bogachiel River, 5, 24, 37
Brager Brothers, 2, 7
Brager, Clarence, 1, 2
Brager, Ella, 1, 2, 25, 42, 45
Brager, Lawrence, 1
Brager, Nancy, 1, 25
Brager, Suzanne, 1

Calawah River, 5
Calawah River Valley, 2, 15, 16; fire risk to, 7
Camp Creek, 7, 9
Clark, Earl, 31, 34, 35, 37, 43, 45
Crown Zellerbach Corp., 13

Drake, Edward "Ed", 14, 45

east wind, 4, 11, 23, 36; described as "gale force", 12; shifts to southwest, 37; strength of, 19
easterly. *See* east wind
Evans, Llewellyn J. "Lew", 15

Fibreboard Products Inc., 7, 43

Floe, Sanford M. "Sandy", 9, 37, 45; calls in Forest Service firefighters, 14
Ford, Dr. Ulric S., 42
Ford, Oliver J. "Ollie", 38
forest fires: characteristics of, 21
Forks: defense of, 16, 32; destroyed buildings in, 35; evacuation of, 18, 24, 30; frontier spirit of, 17; volunteer fire district of, 32
Forks Fire: 1,600-acre blaze controlled, 10; appearance of, 24, 25; August outbreak, 7; blaze jumps Calawah River, 33; crosses Highway 101, 33; financial losses from, 39; litigation from, 43; mopping up of, 37; railroad origins of, 9; rekindled, 11; September sighting, 12; speed of, 15, 26; timber salvage from, 44; turning point of, 36; wind blown, 15
Forks Hospital evacuation, 19
Forks Prairie, 5, 38

Gaydeski, Dixie, 19, 46
Gaydeski, Gary, 19
Gaydeski, Lawrence, 16–19, 46
Gaydeski, Louise, 19
Gunderson Mountain Lookout, 12

Herd, Oscar, 43
Highway 101, 13, 24, 30, 31, 40; fire crosses, 33
Hoh River, 5, 16, 19, 24
Howard, Bill, 16
Huelsdonk, John, 17

Iron Man of the Hoh. *See* Huelsdonk, John

Jensen, Marjorie, 35

MacDonald, John LeRoy "Mac", 13, 35, 37
Mann Gulch fire, 6
Mansfield, Jim, 19, 47
Mansfield, Pat, 19, 47
Mason, Helen, 40
Mason, Jim, 20, 40
Mason, Terry Lin, 40
Maxwell, John, 41
Maxwell, Rue, 41
Merchant, Glen S. "Mickey", 12
Merrill & Ring, 7, 13

North Olympic Peninsula, 5
North Point Lookout Station, 12

Paul, Warren, 45
Port Angeles, 24, 25, 30, 33, 40
Port Angeles and Western Railroad Co., 7; bankruptcy of, 43; deterioration of, 8; fire sighted on, 9; Spruce Division Railroad, 7; steam-powered train, 9

Port Angeles Evening News, 31

Quillayute Naval Air Station, 25

railroad. *See* Port Angeles and Western Railroad Co.
Rayonier Inc., 7, 39

Shearer, Marjorie, 47
Shearer, Reed, 33
Shearer, Ron, 33, 47
Snider Ranger Station, 9, 14, 15, 24
Sol Duc River, 5, 15, 19
Soleduck Ranger District, 5
Spoelstra, John, 27, 35, 40
Spoelstra, Ted, 27, 35, 40, 46
State Division of Forestry, 10, 12, 13, 19
Summer of 1951: drought, 6; fire risk, 6

Thomas, Helen, 28, 47
Thomas, Russ, 28, 41, 47
Tillamook Burn, 6
timber salvage, 44

U.S. Forest Service, 7, 9, 14, 26, 43; fire losses in, 39; salvage sales in, 44
Ulin, Helen, 47
Ulin, Vic, 32, 47

About the Author

Mavis Amundson lived on the Olympic Peninsula from 1987 to 1992. She is the author of *The Lady of the Lake* and the editor of *Sturdy Folk*, both best-selling Peninsula books.

A journalist since 1983, she worked as a reporter and editor for the *Peninsula Daily News* in Port Angeles, and she also worked as a reporter in Seattle and Eugene, Oregon.

She grew up in Seattle and graduated from the University of Washington.

She lives in Seattle with her spouse, George Erb.